Bibliographic information published by the German National Library:

The German National Library lists this publication in the National Bibliography; detailed bibliographic data are available on the Internet at http://dnb.dnb.de .

Imprint:

Print and binding: Books on Demand GmbH, Norderstedt Germany
ISBN: 9783668665088

This book at GRIN:

https://www.grin.com/document/417196

Marcel Strangmueller

A PESTEL Analysis of the company Siemens

GRIN Verlag

TABLE OF CONTENT

Abstract 2

Introduction 3

PESTEL - Analysis 4

Political influence 4

Economical factors 4

Social impact 5

Technology 6

Environmental impact 7

Legal issues 7

Conclusion 8

Reference list 10

ABSTRACT

This essay will give a small insight into Siemens´s history and then will go over to a PESTEL analysis and show how Siemens is doing in the different parts of the PESTEL. The political paragraph will deal with the passed EU law that obligates members to save energy and become more sustainable and how this has an impact on Siemens as they set a greater focus on their sustainable energy sector and get to make more profit out of this. The economical part is going to show how Joe Kaeser as the new CEO did higher the profits from 2012 to 2014 from 4.282 million € up to 5.507 million €. In the social part the SBK as an insurance for their workers as well as the "empowering people award" will be shown, with the "EinDollarBrille" as one the winners that is now used in several developing countries to grant poor people access to cheap, 1$, glasses and a better life. The technological passage will have a short part about how Siemens pushed technology forward from the beginnings of the company on. Furthermore, this passage will focus on the new innovation of "TaxiBots" that pull aircrafts to their starting positions without using their engines and saving fuel that does not have to be used before the flight. Siemens´s impact on the environment is then shown in the next paragraph and how the company helps people in Kenya cleaning the streets, teaching them to live a healthier life and granting them cheap access to electricity and clean, fresh water. After all these positive aspects the last point is the biggest scandal in Siemens´s history in 2006 when they had a huge corruption problem that cost the company 2.5 billion € but made them a role model in the fight against corruption afterwards. The conclusion then sums up the factors and shows that Siemens is a firm that tries to do the morally best and inspire others to do this as well.

INTRODUCTION

Macro environmental factors have always influenced the business in the public and the private sector. This paper will deal with the different factors that have an influence on Siemens and will provide a short overview over the company´s history and will then show the PESTEL factors having an impact on the company and the company having an impact on the factors. "The PESTEL framework is an analytical tool used to identify key drivers of change in the strategic environment. PESTEL analysis includes Political, Economic, Social, Technological, Legal, and Environmental factors." (Johnson 2008). Therefore the essay will deal with the economical part of Siemens, how they are influenced by environmental regulations, their impact on society, the technological advance Siemens brings into the market, the scarcity of resources and a legal issue Siemens had in the earlier past.

The history of Siemens began in 1847 with Werner von Siemens and Johann Georg Halske who founded the Telegraph Construction Company of Siemens and Halske in Berlin. They started of selling Telegraphs and expanding their Telegraph network from Germany into other countries such as Russia. In 1866 Werner von Siemens discovered the dynamo-electric principle and created thereby the basis for heavy-current machinery and technology that is still sold by the company. With this discovery the company started to sell more than just the Telegraphs. In 1903 Siemens and Halske merged with a German joint-stock electricity company until the foundation of the sovereign Siemens AG in 1966 as we know it today. From then on Siemens worked in four main sectors: Healthcare, Energy, Industry and Infrastructure & Cities. Today Siemens´s head is Joe Kaeser who targets a bigger focus on the energy sector as this one is going to be more important in the future. Today Siemens employs around 342.000 employees (March 2015) in branches in around 190 countries but the main office still remains in Germany in Munich and Berlin.

PESTEL - ANALYSIS

POLITICAL INFLUENCE

Siemens as a global player is strongly affected by political decisions made in different countries. But as it is a German company the focus of this paragraph will be on the German and European political decisions and how they have an impact on Siemens. Therefore the policy Germany and the EU are driving at the moment regarding the change of energy production away from nuclear power towards sustainable energies such as wind, solar and tidal power are to mention here. 2007 the members of the EU signed a contract regarding energy production, usage and saving to the year 2020. This contract has terms in it that oblige the member states to higher their coverage of energy usage through sustainable energy sources. Beside this the contract talks about a lowering of energy usage in total by an average of 1,5% per year between the year 2014 and 2020 as well as setting national goals that should be reached by the end of 2020 (Paolo Bertoldi 2007).

This brings focus to the impact those decisions have on Siemens as a company that offers very well developed wind energy technologies for on-shore and off-shore occasions. As the renewable energy sector gets more important the energy supply companies turn their focus onto this more and more. And that is the reason why Siemens is affected in a political way. Siemens supplied 83% of the newly installed wind turbines in central Europe in the last few years and as the green energy sector keeps growing they start making more and more profits out of this department. Some of the major wind parks Siemens supplied with wind turbines are the wind parks "Walney", "Greater Gabbard" and "London Array" (Ludwig Beckers 2010). Siemens has a new order to produce 150 turbines for the offshore wind park "Gemini" in the Netherlands worth around 1.5 billion €.

ECONOMICAL FACTORS

Those new possibilities for orders Siemens has coming in are leading towards the question how Siemens is doing in general economic meanings. How did the company grow over the recent years? What is the value of their shares and are they affected by Germany´s economy?

Due to the fact that Siemens has a new CEO, Joe Kaeser, who set a new policy to higher profits and cut costs Siemens is doing well all in all. Since 2012 Siemens managed to increase their net income from 4.282 million € up to 5.507 million € (Siemens 2014). This is an increase of about 28% in 3 years, which is a big step forward to reaching goals that were set up by Siemens' managing board. The increase of income is composed of two main factors. One factor is already mentioned above, the new incoming orders regarding energy as well as other sector´s well going orders, and the other factor is cutting own costs. The second factor is highly connected to the fact that Siemens abolished a total of nine thousand jobs all over the world from 352,000 in 2012 down to 343,000 at the end of 2014. Joe Kaeser had to stand up for that decision as he was strongly criticized by the press and Siemens employees themselves. In an interview with the German "Abendzeitung-Muenchen" Kaeser stated that it was the hardest decision he had to make. During the re-structuring he puts Siemens through, Kaeser is going to abolish another 4,500 jobs worldwide due to difficulties Siemens has had in the past and the profit plans Kaeser has for the company (Abendzeitung 2015).

The topic of a company's profit leads to the topic of shares and share value. Siemens shares are a good investment for low risk investors. In the past three years the value of its shares rose from 78€ to a maximum of 105€ in April 2015 and are now at a good level of 92€. They had a small down in July to the same value as they had in 2013, 80€, when the German economy had a recession but it went up fast again. (goyax.com 2015)

share value development over the past 5 years (goyax.com)

SOCIAL IMPACT

To show Siemens´s impact on the society and people, there have to be two main topics that have to be named. One is the SBK, which is the insurance Siemens built up for their workers and their families, and the other one are social projects Siemens runs and supports all over the world.

The SBK is a big statutory health insurance that started in the early 20th century. It was created by Siemens and for their workers to provide them with a good healthcare in case of any injuries or other health problems. When it was founded Siemens paid one third of the dues for the insurance and they still do that for their employees. In 1996 they opened the insurance up to their employee´s families and in 1999 SBK decided to let everyone get a member even if they were not employed at Siemens or had any relations to the company. The SBK is the second biggest health insurance in Germany and has over 1 million members. This is one of the contributions Siemens had, and still has, for society.

The other one are projects they support or have supported. One of Siemens´s biggest or most important projects might be the "empowering people award" which was created for innovators, domain experts, users and organisations to share ideas, create a network and give the possibility to discuss problems or new inventions. The award itself is given away to people with unique new inventions to help developing countries to higher living standards. In 2013 the first prize went to the "EinDollarBrille e.V." that created glasses that are affordable for everyone as their total production costs are around 0,8$ and their production only takes around 15 minutes. This helps children in developing countries to attend schools and helps adults to do jobs they were not able to do before. The glasses are in use in countries like Rwanda, Bangladesh, Brazil and Malawi and many more that have a problem with mass poverty.

TECHNOLOGY

Siemens started early to show off the technologies they invented. In 1873 Siemens constructed ship, the "Faraday", to bring out the transatlantic telegraphcable to connect Europe and the US with the fastest communication possibility at that time (Phoocharoon, 2014). 103 words in 15 hours, which was 1 hour faster than their biggest competitor. In 1881 Siemens built the first electric tram, which was then used as role model for any future trams, short, and long train connections (Wilson 1999). Siemens is one of the biggest suppliers in the heavy industry, building technologies and healthcare systems on the planet. As part of their new inventions they developed a new tractor that automatically docks onto airplanes and tows them into take-off position. After multiple tests of those tractors, Lufthansa now regularly uses the "TaxiBots" for their aircrafts. According to the airline the tractors help the airline to save around 11,000 metric tons of fuel each year just at the Frankfurt International Airport. This use could be expanded onto other airports and airlines to save fuel and spare the environment (Lufthansa, 2015).

Environmental impact

The saving of fuel and therefore the protection of natural resources and environment is a good introduction to the environmental part of this essay. As mentioned in the paragraph above Siemens helps already saving fuel on airplanes as part of their sustainability policy. Of course they think of maximising their profit with this help but Siemens has other projects running that are non-profit and help the society to grow sustainable and be aware of what is possible with renewable energy and recycling.

One of these projects are the so called WE!Hubs in Kenya. *WE!Hubs are solar-powered energy stations, which operate independently of electricity grids and can supply green energy and safe drinking water to rural regions without infrastructure* (Siemens 2015). For a small fee people around these stations can rent portable lamps, charge their phones and rent batteries for their own use. The stations offer small jobs for a few people, training programs for locals and a source of income for people living nearby (Siemens, 2015). The fees paid for the offered products at the Hubs are used to pay the workers and expand the business and the surplus of profit is given to other charity organizations in the area. At the moment there are plans to build 5 new WE!Hubs in Kenya around the Lake Victoria, a township in Nairobi near a tea farm in Kericho.

Another project Siemens supports in Kenya is TakaTaka solutions. This organization developed a plan for Nairobi´s waste management. The city's recycling rate is only at about 30%, as a large population cannot afford a proper disposal for their waste. A huge amount of the waste is illegally stored or burnt in the streets, which has a severe impact on the citizens' health. The organization employs around 30 young people and teenagers who would be unemployed otherwise and trained them to collect and recycle the waste of the townships of Nairobi. The regular waste collection improved the living conditions significantly as people are now more aware of the connection of waste in the streets and their personal health condition.

Legal issues

Beside all the bright sides Siemens has it is still an enormous company that is representative in nearly every country all around the world. And as such a big firm it faces problems that come along with the big number of employees. Not every person is always inside the legal boundaries and that causes legal issues. The probably biggest and most known legal issue Siemens had in its past was the corruption scandal of 2006. The scandal consisted of 330 different shady projects that Siemens supported or founded. It was shown that some of the members of the managing board spent money on projects where money should not have been spent and they paid people to make sure that Siemens would get projects and therefore a higher profit. These suspicious short-term high profits alerted the police and they started investigating against the company. All in all after their investigations and with help of the managing committee of Siemens, that deployed extern auditors, it was revealed that Siemens´s managers and other staff had made around 4,300 illegal payments, with a total value of about 1.5 billion €, over several years to get more out of their projects and to be hired for new projects. After the scandal was revealed Siemens had to pay over 2.5 billion euro punishments, after-taxes and professional fees for lawyers and chartered accountants. But as the scandal was over Siemens created an internal anti-corruption policy and is now the German flagship company when it comes to fighting corruption. Siemens therefore made the best out of the situation and created a corruption free company as far as it is known today.

Conclusion

To sum up what can be said about Siemens with the information above is that Siemens as a company is growing in value as it can be seen in the profit they made and how this profit is growing in the next years and grew in the past years. With such an enormous economical influence Siemens has there comes a big social responsibility as global players like Siemens can afford making the world a better place to live and give equality to people that cannot afford a certain standard of living. Siemens meets this social responsibility very well as they support several projects in developing countries to higher the standard of living for poor people there. While they are helping the society in these countries with projects like the one-dollar glasses they help to educate people who might be able to pay for their kids or themselves to have access to a better life. And what comes along with education is the attention given to the environment. Siemens wants to create a more sustainable environment and with projects like the WE!Hubs in Kenya they help poor people to have easy access to electricity, water and work. What directly influences the environment is the education and the recycling work they supported to bring into Nairobi´s townships to clean the streets and higher the awareness of people to the fact that waste in the streets can lead to illnesses and personal indisposition. Beside all these positive factors Siemens has there are of course negative points. One of them are legal issues as nearly every large company has them. Siemens largest scandal was possibly the corruption scandal of 2006 that drew global attention on the company and cost the firm 2.5 billion euro in after-taxes, punishments and other expenses in connection to this case. After that scandal Siemens created a non-corruption environment and is now a role model for other companies. Siemens, however, is a great company and a good start for everyone whose goal is to work in the industry and have a social responsible employer.

Reference list

Abendzeitung, M. (2015) *Nach extremen Konzernumbau: Siemens-Chef Kaeser verteidigt Stellenabbau - Abendzeitung München*. Available at: http://www.abendzeitung-muenchen.de/inhalt.nach-extremen-konzernumbau-siemens-chef-kaeser-verteidigt-stellenabbau.5db1538d-968e-4042-94cf-4e72b851e89f.html (Accessed: 3 November 2015).

Aschenbrenner, N. (2014) *Electric mobility: New aircraft tractors*. Available at: http://www.siemens.com/innovation/en/home/pictures-of-the-future/mobility-and-motors/electromobility-aircraft-tractors.html (Accessed: 3 November 2015).

Beckers, L. (2010) 'Bautechnik aktuell: Bautechnik 5/2010', *Bautechnik*, , pp. 308–310. doi: 10.1002/bate.201090051.

Berlin and Niebüll (2012) *Energiewende*. Available at: http://www.economist.com/node/21559667 (Accessed: 25 October 2015).

Bertoldi, P. (2007) *Electricity Consumption and Efficiency Trends in the Enlarged European Union*. Available at: http://www.qualenergia.it/sites/default/files/articolo-doc/Electricity%20Consumption%20in%20UE.pdf (Accessed: 4 November 2015).

Fichter, D. F. (2015) *DIW Berlin: Konjunkturbarometer vom 28. Oktober 2015*. Available at: http://www.diw.de/de/diw_02.c.102177.de/forschung_beratung/daten/konjunkturbarometer/konjunkturbarometer_vom_26_maerz_2015.html (Accessed: 3 November 2015).

IWR (2015) *Siemens baut weiteren großen Offshore-Windpark*. Available at: http://www.iwr.de/news.php?id=30013 (Accessed: 3 November 2015).

Johnson, G. (2008) *PESTEL analysis - what is it? Definition, examples and more*. Available at: https://www.kbmanage.com/concept/pestel-analysis (Accessed: 23 October 2015).

Leifiphyik (no date) *Physikportal*. Available at: http://www.leifiphysik.de/themenbereiche/elektromagnetische-induktion/dynamoelektrisches-prinzip (Accessed: 3 November 2015).

Leyendecker, H. (2011) *Siemens: Korruptionsaffäre – 'das ist wie bei der mafia'*. Available at: http://www.sueddeutsche.de/wirtschaft/siemens-korruptionsaffaere-das-ist-wie-bei-der-mafia-1.1046507 (Accessed: 3 November 2015).

Lufthansa (2015) *Lufthansa rolls out innovative TaxiBots for real flight operations*. Available at: http://www.travelandtourworld.com/news/article/lufthansa-rolls-innovative-taxibots-real-flight-operations/ (Accessed: 1 November 2015).

Phoocharoon, P. (2014) *Werner Von Siemens: Transpirational Leader on Leading the Revolution*. Available at: http://icehm.org/siteadmin/upload/8548ED0814077.pdf (Accessed: 4 November 2015).

Richtlinien Europäisches Parlament (2006) Available at: http://eur-lex.europa.eu/LexUriServ/LexUriServ.do?uri=OJ:L:2009:140:0016:0062:de:PDF (Accessed: 26 October 2015).

SBK (no date) *Alle Leistungen*. Available at: https://www.sbk.org/leistungen/alle-leistungen/ (Accessed: 3 November 2015).

SBK (no date) *Alle Leistungen*. Available at: https://www.sbk.org/leistungen/alle-leistungen/ (Accessed: 3 November 2015).

Siemens (2014) *Annual Siemens Report*. Available at: http://www.siemens.com/annual/14/en/download/pdf/Siemens_AR2014.pdf (Accessed: 3 November 2015).

Siemens (2014) *History*. Available at: http://www.siemens.com/about/en/history.htm (Accessed: 19 October 2015).

Siemens (2015) *Siemens Projects*. Available at: http://www.siemens-stiftung.org/de/projekte/ (Accessed: 29 October 2015).

Siemens, S. (2015) *Empowering-people*. Available at: http://www.empowering-people-network.siemens-stiftung.org/index.php?id=37 (Accessed: 3 November 2015).

Stiftung, S. (2015) *Takataka-solution*. Available at: http://www.siemens-stiftung.org/en/projects/takataka-solutions/ (Accessed: 3 November 2015).

Unknown (2014) *Siemens erhält Milliardenauftrag für Windräder*. Available at: http://www.abendblatt.de/wirtschaft/article128066696/Siemens-erhaelt-Milliardenauftrag-fuer-Windraeder.html (Accessed: 28 October 2015).

Wilson, R. and Budd, D. (1999) *The Melbourne Tram book: 3rd edition*. Available at: https://books.google.de/books?hl=de&lr=&id=QzBtBgAAQBAJ&oi=fnd&pg=PA3&dq=Siemens+first+electric+tram&ots=ikFSSfsm7I&sig=J7UTHPdsDXWWRWF4Nk-

QdgsmYhs#v=onepage&q=Siemens%20first%20electric%20tram&f=false (Accessed: 4 November 2015).

goyax (2015) *Siemens Aktie - Aktienkurs | Kurs | DE0007236101 | 723610 @ GOYAX*. Available at: http://www.goyax.de/siemens-Aktie (Accessed: 3 November 2015).